ABBREVIATED VERSION

Toward a Universal Radio Frequency System for SPECIAL OPERATIONS FORCES

Committee on Universal Radio Frequency System for Special Operations Forces

Standing Committee on Research, Development, and Acquisition Options for U.S. Special Operations Command

Division on Engineering and Physical Sciences

NATIONAL RESEARCH COUNCIL
OF THE NATIONAL ACADEMIES

THE NATIONAL ACADEMIES PRESS
Washington, D.C.
www.nap.edu

THE NATIONAL ACADEMIES PRESS 500 Fifth Street, N.W. Washington, DC 20001

NOTICE: The project that is the subject of this report was approved by the Governing Board of the National Research Council, whose members are drawn from the councils of the National Academy of Sciences, the National Academy of Engineering, and the Institute of Medicine. The members of the committee responsible for the report were chosen for their special competences and with regard for appropriate balance.

This is a report of work supported by Contract H92222-07-D-0014, DO #3, between U.S. Special Operations Command and the National Academy of Sciences. Any opinions, findings, conclusions, or recommendations expressed in this publication are those of the author(s) and do not necessarily reflect the view of the organizations or agencies that provided support for the project.

International Standard Book Number 13: 978-0-309-13242-8
International Standard Book Number 10: 0-309-13242-8

Limited copies are available from:

Division on Engineering and Physical Sciences
National Research Council
500 Fifth Street, N.W.
Washington, DC 20001
(202) 334-3111

Additional copies available from:

The National Academies Press
500 Fifth Street, N.W.
Lockbox 285
Washington, DC 20055
(800) 624-6242 or (202) 334-3313
(in the Washington metropolitan area)
http://www.nap.edu

Printed in the United States of America

THE NATIONAL ACADEMIES

Advisers to the Nation on Science, Engineering, and Medicine

The **National Academy of Sciences** is a private, nonprofit, self-perpetuating society of distinguished scholars engaged in scientific and engineering research, dedicated to the furtherance of science and technology and to their use for the general welfare. Upon the authority of the charter granted to it by the Congress in 1863, the Academy has a mandate that requires it to advise the federal government on scientific and technical matters. Dr. Ralph J. Cicerone is president of the National Academy of Sciences.

The **National Academy of Engineering** was established in 1964, under the charter of the National Academy of Sciences, as a parallel organization of outstanding engineers. It is autonomous in its administration and in the selection of its members, sharing with the National Academy of Sciences the responsibility for advising the federal government. The National Academy of Engineering also sponsors engineering programs aimed at meeting national needs, encourages education and research, and recognizes the superior achievements of engineers. Dr. Charles M. Vest is president of the National Academy of Engineering.

The **Institute of Medicine** was established in 1970 by the National Academy of Sciences to secure the services of eminent members of appropriate professions in the examination of policy matters pertaining to the health of the public. The Institute acts under the responsibility given to the National Academy of Sciences by its congressional charter to be an adviser to the federal government and, upon its own initiative, to identify issues of medical care, research, and education. Dr. Harvey V. Fineberg is president of the Institute of Medicine.

The **National Research Council** was organized by the National Academy of Sciences in 1916 to associate the broad community of science and technology with the Academy's purposes of furthering knowledge and advising the federal government. Functioning in accordance with general policies determined by the Academy, the Council has become the principal operating agency of both the National Academy of Sciences and the National Academy of Engineering in providing services to the government, the public, and the scientific and engineering communities. The Council is administered jointly by both Academies and the Institute of Medicine. Dr. Ralph J. Cicerone and Dr. Charles M. Vest are chair and vice chair, respectively, of the National Research Council.

www.national-academies.org

COMMITTEE ON UNIVERSAL RADIO FREQUENCY SYSTEM FOR SPECIAL OPERATIONS FORCES

LAWRENCE DELANEY, *Chair*, Titan Corporation (retired)
BRIAN AGEE, B3 Advanced Communications Systems
MARK BUCKNER, Oak Ridge National Laboratory
R. MICHAEL BUEHRER, Virginia Polytechnic Institute
JOHN CAFARELLA, Independent Consultant
PHILIP DICKINSON, Independent Consultant
JAMES FREEBERSYSER, BBN Technologies
RITA GONZALES, Sandia National Laboratories
ANDREW IVERS, L-3 Linkabit
PETER KIND, Institute for Defense Analyses
GARY MINDEN, University of Kansas
JOSEPH MITOLA, Stevens Institute of Technology
ROBERT NOWAK, Defense Advanced Research Projects Agency (retired)
GLEN ROUSSOS, L-3 Communications
ROBERT TINGLEY, Charles Stark Draper Laboratory

Staff

CARTER FORD, Study Director
NORMAN HALLER, Consultant
KAMARA BROWN, Research Associate
ENITA WILLIAMS, Research Associate (through October 2008)
URRIKKA WOODS, Program Associate

STANDING COMMITTEE ON RESEARCH, DEVELOPMENT, AND ACQUISITION OPTIONS FOR U.S. SPECIAL OPERATIONS COMMAND

JACQUES GANSLER, *Chair*, University of Maryland
A. MICHAEL ANDREWS, L-3 Communications
KENNETH BOWRA, Oak Ridge National Laboratory
LAWRENCE DELANEY, Titan Corporation (retired)
CHIP ELLIOTT, BBN Technologies
MICHAEL HOPMEIER, Unconventional Concepts, Inc.
JOSEPH MANCUSI, Narus Government Solutions
RANDALL MURCH, Virginia Polytechnic Institute
WILLIAM NEAL, The MITRE Corporation
ROBERT NOWAK, Defense Advanced Research Projects Agency (retired)
ALTON ROMIG, JR., Sandia National Laboratories
PAUL ROSENSTRACH, The Charles Stark Draper Laboratory
JULIE J.C.H. RYAN, George Washington University
HARRY SCHULTE, Raytheon Missile Systems
ANNETTE SOBEL, University of Missouri
JOHN SOMMERER, Johns Hopkins University Applied Physics Laboratory
DWIGHT STREIT, Northrop Grumman Corporation
DAVID WHELAN, The Boeing Company

Staff

MICHAEL CLARKE, Director
CARTER FORD, Program Officer
NORMAN HALLER, Consultant
DANIEL TALMAGE, JR., Program Officer
KAMARA BROWN, Research Associate
URRIKKA WOODS, Program Associate
MARGUERITE SCHNEIDER, Administrative Coordinator

Preface

The U.S. Special Operations Command (SOCOM) was formed in response to the failed rescue attempt in 1980 of American hostages held by Iran. Among its key responsibilities, SOCOM plans and synchronizes operations against terrorist networks. Special operations forces (SOF) often operate alone in austere environments with only the items they can carry, which makes equipment size, weight, and power needs especially important. Specialized radios and supporting equipment must be carried by the teams for their radio-frequency (RF) operations. As warfighting demands on SOCOM have intensified, SOCOM's needs for significantly improved radio-frequency (RF) systems have increased.

STATEMENT OF TASK

In September 2007, SOCOM's acquisition executive discussed with the National Research Council's (NRC's) SOCOM Standing Committee whether the state of the art could produce a suitable handheld RF system with universal capabilities for SOF missions. The NRC agreed to conduct a technical study and negotiated the statement of task (SOT) with SOCOM to address the following areas:

- Examine the current state of the art for both handheld and manpackable platform-mounted radio frequency (RF) systems available or in development by industry, national and service laboratories, and university research establishments;
- Based on special operations forces (SOF)-unique RF requirements, determine which frequencies could be provided by handheld systems. The handheld system should provide SOF capabilities such as command and control; situational awareness and tracking; navigation and geolocation; hostile force tagging, tracking, and locating; signals intelligence; SOF blue force tracking; communications; and counter improvised explosive devices;
- Determine if such a system could be deployed in a reasonable time period at a technology readiness level and utility of use to SOF;
- Evaluate methods of using extant systems; and
- Issue a report that provides recommendations to address the above.[1,2]

---

[1]In this report the committee used the term "RF system" to include the radio, display, power amplifier, antenna, and power source. Looking ahead, a "universal RF system" would enable a robust complement of capabilities and attributes of the type envisioned for future operations by SOF. The committee's ultimate vision of the universal RF system includes a "modular" handheld radio, a sophisticated display to support several capabilities (e.g., situational awareness), a power amplifier, antennas with up- and downconverters to cover microwave frequency bands, and required power sources. The term "modular" refers to the basic radio building block, to which accessories can be added as needed to realize the full range of universal RF capabilities.

THE COMMITTEE APPROACH

The Committee on Universal Radio Frequency System for Special Operations Forces was formed to conduct the study (see Appendix A for short biographies of the committee members). One of the committee's early challenges was to determine how it would fulfill the SOT. The committee met with the sponsor in June, July, and August of 2008 to receive more information on matters relevant to the tasking—for example, the type of information it would need to understand fully the command's needs for RF systems. The committee also met and corresponded with outside experts well versed in radio technologies (see list of meetings in Appendix B).

Based on its meetings and the committee's own expertise, it was clear that the task was intended to go beyond merely studying "which frequencies could be provided by handheld systems." Rather, a comprehensive RF-system approach would be required to fulfill the intent of the study. In bounding its approach, however, the committee notes that a separate NRC entity, the Committee on Sensing and Communications Capabilities for Special Operations Forces, is studying sensor technologies for SOCOM that may have unique communications requirements and systems. That report is scheduled for delivery to SOCOM in the summer of 2009.[3]

During the course of the study, the committee developed a list of capabilities and associated attributes for universal RF systems that could meet the mission needs of the sponsor. These capabilities and attributes were based on information the committee received from military and commercial sources and on the committee's expertise in this field. In this report, the capabilities and attributes are analyzed with regard to their enabling technologies and, most importantly, the time frames for their likely availability.[4] Three time frames were considered: the near term (2009-2011), the medium term (2011-2013), and the far term (beyond 2013). The committee focused especially on the near term, recognizing that SOCOM desired a significant advance in capability within approximately 2 years. Finally, the committee estimated time frames for either modifying existing RF systems or developing advanced universal RF-system design concepts. The committee also offered recommendations concerning various aspects of the acquisition by SOCOM of such systems.

The months between the committee's last meeting and the publication of the report were spent preparing the draft manuscript, reviewing and responding to the external peer review comments, editing the report, and conducting the required security/public release review necessary to produce this version of the report that does not disclose information as described in 5 U.S.C. 552(b). It was mutually determined by SOCOM and the NRC

---

[2]Readers should be aware that in discussions SOCOM, which is fully versed in knowledge of the Joint Tactical Radio System (JTRS) system development process and its goals, specifically asked the committee to consider and assess RF system opportunities other than the JTRS.

[3]The ad hoc committee notes that the scope of a separate NRC study, *Sensing and Supporting Communications Capabilities for Special Operations Forces: Report 1*, is complementary to the scope of this report.

[4]Early in the committee's deliberations the matter of delineating specific technology readiness levels was raised. Convinced that the technology readiness definitions used by the Department of Defense (DOD) are too formulaic to be broadly applicable across the radio technologies considered in this report, the committee decided to discuss the capabilities in terms of time frames when different levels of enabling technology would be ready for a system to be of utility of use to SOF. General descriptions of the technology levels are given here for the near term (minimal modifications to existing products), the medium term (extensive modifications as well as new development), and the far term (ground up development). The DOD numerical system of technology readiness level (TRL), which extends from 1 to 9, would correspond, roughly, to the committee's system of near term (TRL 7-9), medium term (TRL 4-6), and long term (TRL 1-3).

that the full report contained information as described in 5 U.S.C. (b) and therefore could not be released to the public in its entirety.

The committee thanks all who contributed to this effort, including the individual committee members who volunteered considerable time and energy, the supporting staff of the National Academies, and the many outside experts who provided necessary information to inform committee deliberations. Finally, the committee greatly appreciates the opportunities to interact with the knowledgeable SOCOM representatives, who shed much light on how their forces operate and emphasized the importance of this work to their warfighters.

Lawrence Delaney, *Chair*
Committee on Universal Radio Frequency
System for Special Operations Forces

Acknowledgment of Reviewers

This report has been reviewed in draft form by individuals chosen for their diverse perspectives and technical expertise, in accordance with procedures approved by the National Research Council's (NRC's) Report Review Committee. The purpose of this independent review is to provide candid and critical comments that will assist the institution in making its published report as sound as possible and to ensure that the report meets institutional standards for objectivity, evidence, and responsiveness to the study charge. The review comments and draft manuscript remain confidential to protect the integrity of the deliberative process. We wish to thank the following individuals for their review of this report:

David Borth, Motorola, Inc.,
William Bridges, California Institute of Technology (emeritus),
Gary Brown, Virginia Polytechnic Institute and State University,
Joel Engel, JSE Consulting,
William Neal, The MITRE Corporation,
Harry Schulte, Raytheon Missile Systems,
John Stenbit, TRW, Inc. (retired), and
George Swenson, University of Illinois (emeritus).

Although the reviewers listed above have provided many constructive comments and suggestions, they were not asked to endorse the conclusions or recommendations nor did they see the final draft of the report before its release. The review of this report was overseen by R. Stephen Berry, University of Chicago. Appointed by the NRC, he was responsible for making certain that an independent examination of this report was carried out in accordance with institutional procedures and that all review comments were carefully considered. Responsibility for the final content of this report rests entirely with the authoring committee and the institution.

Contents

Toward a Universal Radio Frequency System for Special Operations Forces: Abbreviated Version

OVERARCHING NEEDS AND RESPONSE

Two overarching needs of the U.S. Special Operations Command (SOCOM) drove this study by the Committee on Universal Radio Frequency System for Special Operations Forces: (1) reduce the quantity and weight of equipment carried on a mission and (2) add future capabilities quickly and efficiently. The committee's overarching response concerned what SOCOM could expect to be possible over time, especially in the near term (2009-2011). The other time frames were defined as the medium term (2011-2013) and the far term (beyond 2013).

The radio frequency (RF) systems now used by special operations forces (SOF) consist of a collection of special-purpose systems that can satisfy many current mission needs. However, that collection does not possess the full range of capabilities and attributes needed by SOF today and in the future, and it is very heavy.

The committee developed approximate near-, medium-, and far-term acquisition time frames that could permit deployment of improved, universal RF systems. In the near term, existing handheld RF systems could potentially replace the manpackable systems in the inventory for many applications. In the medium term, the bandwidth and processing speeds of existing RF systems could be extended, and the addition of antennas with increased gain or active processing, preamplifiers, and downconverters could provide coverage of most or all of the frequencies of interest.

The far-term period would start at the same time as the medium-term period but last roughly twice as long. That duration, and the large number of potential future capabilities and attributes that could be enabled by the technologies identified in this report, suggest guidelines that would support robust universal RF-system solutions with markedly reduced size, weight, and power consumption. These include (1) employing a basic modular radio as a standard building block; (2) simplifying requirements wherever possible (e.g., minimizing the number of signals in space); (3) planning to insert application-specific integrated circuit (ASIC) technology to reduce size and weight and power consumption; and (4) adopting standard smart batteries and supplemental power sources.

CURRENT STATE OF THE ART FOR RADIO FREQUENCY SYSTEMS

Introduction

In accordance with the statement of task (SOT), the committee examined the state of the art for both handheld and manpackable platform-mounted radio frequency (RF) systems that are available or in development by industry, the national and service laboratories, and university research establishments in the following areas: command and control; situational awareness and tracking; navigation and geolocation; hostile force tagging, tracking, and locating; signals intelligence; SOF blue force tracking; communications; and countering improvised explosive devices (IEDs). It also examined five areas not specifically called out in the project scope that it believes are critical support functions without which the goals of a universal radio frequency system (URFS) cannot be attained—namely, information assurance, which includes TRANSEC/COMSEC, antitamper and low probability of intercept/low probability of detection (LPI/LPD); size, weight, and power (SWaP); usability; interoperability; and, sustainability. For each area, the committee considers the systems that are fielded by SOF (roughly TRL-7 and higher), the systems available (TRL-5 and greater) or fielded by other parts of the Department of Defense (DOD), and the systems under development as an emerging technology (TRL-4 and below). For the sake of brevity, the committee has not attempted to exhaustively cover every product, system, or program that falls into each category; the systems covered are representative of the current state of the art in each of the three technology time frames.

Command and Control, Including Situational Awareness and SOF Blue Force Tracking

While the statement of task separates command and control (C2) from situational awareness and tracking and blue force tracking, in this chapter these areas are merged, because proper C2 implies situational awareness where the own-force aspect is obtained via SOF blue force tracking. Information on enemy forces needed for situational awareness is covered in the sections on hostile force tagging, tracking, and location and on signals intelligence.

Fielded Systems

Force XXI Battle Command, Brigade and Below, for Blue Force Tracking[1]

The Force XXI Battle Command Brigade and Below (FBCB2) system forms the principal digital command and control system for the Army at brigade level and below. All FBCB2 systems are interconnected through a communications infrastructure called the Tactical Internet to exchange situational awareness data and conduct C2. FBCB2 is interoperable with the maneuver control system, the All Source Analysis System, the Advanced Field Artillery Tactical Data System, the Air and Missile Defense Work Station, and the Combat Service Support Control System. Blue force tracking employs L-band satellite communication links that have proved to be reliable over long distances and in mountainous terrain. Georeference data displayed includes enemy positions, friendly positions, and hazards and obstacles. The prime contactor for FBCB2 is Northrop Grumman Mission Systems (Dominguez Hill, California).

Command Post of the Future[2]

The Command Post of the Future (CPOF) is an executive-level decision support system providing situational awareness and collaborative tools for tactical decision making, planning, rehearsal, and execution management from corps to battalion level. CPOF supports visualization, information analysis, and collaboration in a single, integrated environment. CPOF operators interactively collaborate, sharing thoughts, workspace, and plans to analyze information and to evaluate courses of action with real-time feedback for an immediate and comprehensive view of

---

[1]For more information on FBCB2, see http://peoc3t.monmouth.army.mil/fbcb2/about.html. Last accessed September 29, 2008.

[2]For more information on CPOF, see http://peoc3t.monmouth.army.mil/battlecommand/bc_CPOF.html. Last accessed September 29, 2008.

the battlefield. CPOF is designed to enable deep cohesion-of-thought processes between the commander and his staff. Users are able to selectively and dynamically generate and transmit their evolving analysis, plans, and execution. General Dynamics is the prime contractor for CPOF.

Joint Tactical Common Operational Picture Workstation[3]

Joint Tactical Common Operational Picture (COP) Workstation (JTCW), formally known as the command and control personal computer (C2PC), is an integrated build of software applications including a Windows-based operating system (OS) designed to facilitate military C2 functions by improving situational awareness and enhancing the commander's operational and tactical decision-making capability. JTCW components consist of the JTCW Client and Gateway as the core software application, Marine Corps and Joint third-party application extension, government off-the-shelf and commercial off-the-shelf applications, and the Microsoft Windows OS. The JTCW is the primary point of entry for the common tactical picture, in which users are able to view map data, manage and track data, develop and distribute overlays and graphics, exchange message traffic, plan and distribute route information, and conduct C2 planning during real-time operations. The C2PC prime contactor is Northrop Grumman Mission Systems (San Diego, California).

FalconView

FalconView is a mapping system created by the Georgia Tech Research Institute[4] for the Windows[5] family of operating systems. It displays various types of maps and geographically referenced overlays. Many types of maps are supported, but the primary ones of interest to most users are aeronautical charts, satellite images, and elevation maps. FalconView also supports a large number of overlay types that can be displayed over any map background. The current overlay set is targeted at military mission planning users and is oriented to aviators and aviation support personnel. FalconView is an integral part of the portable flight planning software.[6] This software suite includes FalconView, combat flight planning software, combat weapon delivery software, Combat Air Drop Planning Software, and several other software packages built by various software contractors.

DARPA TIGR System

Tactical Ground Reporting (TIGR) is a multimedia reporting system for soldiers at the patrol level, allowing users to collect and share information to improve situational awareness and to facilitate collaboration and information analysis among junior officers. With its geospatial user interface, TIGR is particularly suited to counterinsurgency operations and enables collection and dissemination of fine-grained intelligence on people, places, and insurgent activity. Focused on users at the company level and below, TIGR complements existing reporting systems that focus on the needs of users at the battalion or brigade level and above. TIGR's graphical, map-

---

[3]For more information on C2PC, also known as JTCW, see http://hqinet001.hqmc.usmc.mil/p&r/Concepts/2005/PDF/Ch3PDFs/CP05%20Ch3P1%20CEP%20pg%20143_Command%20and%20Control%20Personal%20Computer.pdf. Last accessed September 29, 2008.

[4]For more information on the Georgia Tech Research Institute, see http://www.gtri.gatech.edu/. Last accessed September 29, 2008

[5]For more information on Microsoft Windows, see http://www.microsoft.com/Windows/default.aspx. Last accessed September 29, 2008.

[6]For more information on PFPS, visit http://www.tybrin.com/DisplayArticle.aspx?tblName=tblServices&articleID=37. Last accessed September 29, 2008.

referenced user interface is highly intuitive and allows multimedia data such as voice recordings, digital photos, and GPS tracks to be easily collected and searched. The system also uses a state-of-the-art data distribution architecture to minimize load on the tactical networks while allowing digital imagery and other multimedia data to be rapidly exchanged. TIGR was developed by the Defense Advanced Research Projects Agency (DARPA) under an extremely aggressive schedule. The system was first introduced to users during the predeployment training exercise at Fort Hood in the spring of 2006. The system is currently in experimental use in Iraq.[7]

Available

DARPA Active Templates

The DARPA Active Templates program was established to develop a scalable, simple, distributed software infrastructure for mission planning and execution, in essence a spreadsheet for planning, information monitoring, and execution replanning. This effort addressed the concept of spreadsheets for planning by developing a suite of forms-based planning tools. The objective was to enable users to create and modify forms with sharable information elements to support real-time collaboration, and a core technology was implemented to facilitate collaborative form development. The resulting technology was then used as a foundation for a number of demonstration applications, including weather report visualization, command logs, and, more important, a general form-building application called CommandLink. CommandLink provides a simple, intuitive tool for users to support planning with unique features that enable real-time collaboration, reusability of information elements, and connectivity to external information sources.[8]

Emerging Technology

Unified Battle Command

There is ongoing discussion among the program offices of the current FBCB2 and Future Combat Systems, U.S. Army C2 systems, and the current U.S. Marine Corps C2 system about merging these systems at some point in the future into a unified battle command system. The data and timeline for such a merger are still to be determined, but the committee includes them because they will be relevant to a future URFS.

DARPA Deep Green

Deep Green is a next-generation, commander-centered battle command and decision support technology that interleaves anticipatory planning with adaptive execution to help the commander think ahead, identify when a plan is going awry, and prepare options before they are needed. The Deep Green concept is an innovative approach to using simulation to support ongoing military operations while they are being conducted. By using information acquired from the ongoing operation rather than assumptions made during the planning phase, commanders and their staffs can make more informed choices. The basic system architecture comprises the Commander's Associate (with three subcomponents: the Sketch to Plan, Automated Options Generation, and the Sketch to Decide), Blitzkrieg, and Crystal Ball, as shown in Figure 1.

---

[7]For more information on TIGR, visit http://www.darpa.mil/ipto/programs/assist/assist_tigr.asp. Last accessed September 29, 2008.

[8]For more information on CommandLink, visit http://www.stormingmedia.us/97/9726/A972634.html. Last accessed September 29, 2008.

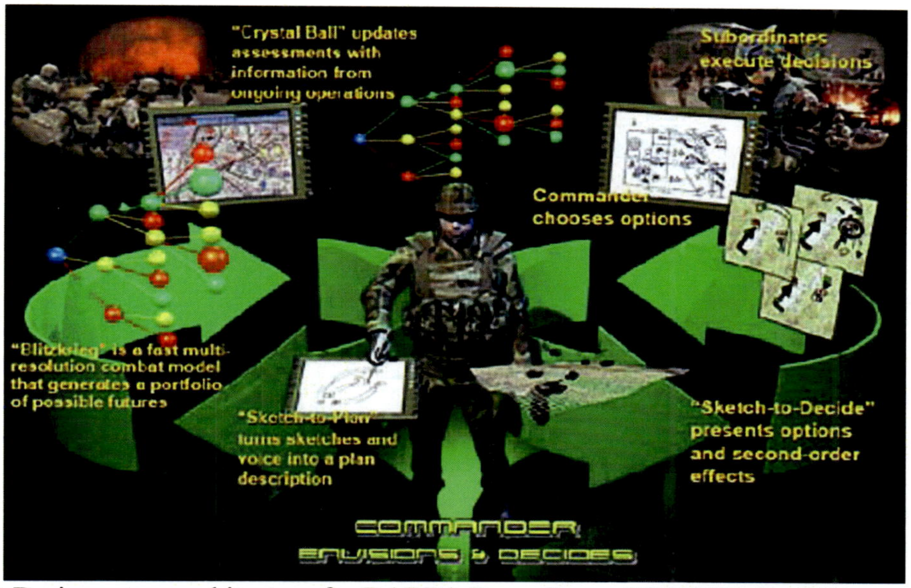

FIGURE 1 Basic system architecture for DARPA Deep Green. SOURCE: DARPA.

Navigation and GeoLocation

Navigation and geolocation in the context of this report are the ability to know one's own position using external (e.g., GPS) or organic (e.g., inertial navigation system (INS)) methods.

Fielded

Precision Lightweight GPS Receiver

The precision lightweight GPS receiver (PLGR) is a handheld, single-frequency military GPS receiver that incorporates the Selective Availability Anti-Spoofing Module (SAASM) to access the encrypted P(Y) code GPS signal. More than 165,000 PLGRs were procured worldwide from the time the system was first introduced, in January 1994, through 2004.

Defense Advanced GPS Receiver

The defense advanced GPS receiver (DAGR) is used by the DOD and select foreign military services. It is a military-grade, dual-frequency receiver that incorporates SAASM. Manufactured by Rockwell Collins, the DAGR entered production in March 2004, with the 40,000th unit delivered in September 2005. It was estimated by *Defense Industry Daily* that by the end of 2006 the U.S. Army and various allies around the world had ordered almost 125,000 units.[9] DAGR features include:

- Two-frequency (L1/L2) operation provides calibration of the ionosphere. Single-frequency operation has a nominal horizontal location accuracy of 10.2 m compared to dual-frequency operation, which offers 6.6 m nominal accuracy. Neither quoted accuracy has any significant geometrical dilution of precision.

---

[9]For more information, visit "$82.7M more for DAGR GPS receivers" at http://www.defenseindustrydaily.com/827m-more-for-dagr-gps-receivers-02829/#more. Last accessed September 29, 2008.

- SAASM uses the secure (encrypted) P(Y) code as well as the coarse/acquisition code (C/A) used by commercial receivers. Besides having improved antijam processing gain (approximately 6 dB more than the C/A receiver), the secure code cannot be spoofed, as can the C/A receiver because of its known and repeated code. SAASM receivers can also maintain high location accuracy when selective availability (SA) is in operation; SA was turned off in May 2000 and most likely will not be turned on again.
- DAGR can use local area differential GPS or wide-area GPS enhancement (WAGE). WAGE is a method to increase the horizontal accuracy of the GPS encrypted P(Y) code by adding range correction data to the satellite broadcast navigation message. The horizontal accuracy using WAGE is approximately 5 m.

Commercial GPS

Commercial GPS receivers can only use the C/A code, which is transmitted on a single frequency. Selective availability had limited the location accuracy of C/A receivers, but it is no longer active. The location accuracy in benign environments is about the same as that of a single-frequency SAASM receiver. Current commercial GPS receivers are highly integrated and inexpensive in large part because they do not require SAASM. They are easily integrated into various prices of equipment such as handheld radios, handheld navigation systems, and cell phones. Differential GPS is widely available; there are local systems as well as a worldwide system, similar to the WAGE, that provides local corrections for many sources of location error. John Deere claims accuracies of 10 cm for controlling farming equipment and for crop mapping.[10] Any performance figures from the commercial manufacturers of GPS should be weighted very highly by the caveat that they are not designed to operate in the unfriendly environments in which SOF operate.

Available

SAASM Chip Sets

The critical component of a SAASM chip set is the key data processor (KDP). Sandia National Laboratories, the KDP developer for the government, has provided several generations of government-furnished equipment (GFE), which have evolved from a five-chip set (KDP-I) to a three-chip set (KDP-II) to a single chip (KDP-III, KDP-IV) that can be instantiated into a single system-on-a-chip SAASM GPS receiver. The KDP-II's current use dates from the mid-1990s and consumes considerable power. All currently fielded SAASM GPS receivers are constructed as multichip modules and include the SAASM developer receiver chips and the GFE KDP-II chips within the multichip module.[11] There are several advantages to creating a fully integrated, single monolithic KDP. Some of those benefits include providing SAASM developers with the ability to create a fully integrated system-on-a-chip single-chip SAASM receiver that provides increased security (no exposed classified interconnects) and reduces costs (elimination of antitamper coating and complex multichip module fabrication), reduces size and weight, and uses an industry-standard trusted integrated circuit fabrication process. First under Navy funding and subsequently under Army funding, one of the SAASM developers began an effort during 2005 to produce the first fully integrated SAASM GPS. In this effort, Sandia was to produce the KDP-III as an IP core that could be integrated at a Trusted Foundry with the company's receiver and processor cores. This development is in 130-nm technology and can be expected to produce

[10]For more information on John Deere farming equipment, visit http://www.progressiveengineer.com/pewebbackissues2005/PEWeb%2060%20Mar05-2/Deere.htm. Last accessed September 29, 2008.

[11]For more information on SAASM, see SAASM System Specification SAASM-ss_gps-001a.

equipment during 2009. Subsequently, a second SAASM developer has begun a development at a Trusted Foundry in 90-nm technology, with Sandia supplying the KDP-IV. This effort should produce equipment during 2010. The applications targeted for this high level of integration are guidance for artillery shells and missiles, land mines, unattended ground sensors, and handheld radios.

M-Code

GPS M-code receivers will have all SAASM capabilities—that is, correction for situational awareness and processing of the P(Y) and C/A codes. All M-code receivers will be fully integrated simply because that is the emerging state of the art of SAASM GPS receivers. The most significant changes effected for M-code equipment are these: (1) M-code satellites will provide a factor of 100 (20-dB) higher transmit power from the satellite and (2) the M-code signal in space is a 5 Mchip per second code rate, wherein each chip waveform is Manchester encoded. This produces a null in the M-code signal spectrum at center frequency, enabling jamming of the C/A code to deny its use to unfriendly forces. The manufacturers of M-code equipment will handle security rather than the government supplying hardware and keying receivers.

At present, it appears that launch of the initial M-code satellites has been delayed, and it is not clear when manufacturers will introduce M-code receivers.

Emerging Technology

DARPA Robust Surface Navigation[12]

The DARPA robust surface navigation (RSN) program will provide DOD with the ability to geolocate and navigate effectively when GPS is unavailable due to hostile action (e.g., jamming) or blockage by structures and foliage. The RSN program will develop the procedures and technologies for geolocation of stationary assets and navigation of mobile platforms by exploiting signals of opportunity and/or specialized signals from satellite, airborne, and terrestrial assets. The use of widely available, powerful, and economically important (and thus dependable) signals of opportunity will provide a robust non-GPS capability. Signals of opportunity can also be augmented when necessary by purpose-deployed, signal-emitting beacons. RSN will use the greater strength and diversity of these opportunistic and intentional signals to provide coverage when GPS is denied due to lack of penetration, when severe multipath is a problem, or when GPS is jammed or denied globally. ArgonST (Fairfax, Virginia) is the prime contractor.

DARPA Precision Inertial Navigation Program[13]

Military navigation systems use updates from GPS satellites to enhance the INS's knowledge of their current position. However, GPS transmission is vulnerable to jamming, and signal reception is difficult or impossible in certain geographic conditions (underwater, in urban or natural canyons, underground, etc.). Without GPS updates, INS positional accuracy drifts with time at a few miles per hour (referred to as the drift rate of the navigation system). The precision inertial navigation systems (PINS) program seeks to use ultracold atom interferometers as an alternative to GPS position updates. Advances in atomic physics over the past two decades have allowed scientists exquisite control over the external quantum states of atoms, including the deliberate production of matter waves from ultracold atoms. This has allowed the development of

[12]For more information on DARPA's RSN, see http://www.darpa.mil/sto/space/rsn.html. Last accessed September 29, 2008.

[13]For more information on DARPA's PINS, see http://www.darpa.mil/dso/thrusts/physci/newphys/pins/index.htm. Last accessed September 29, 2008.

matter-wave interferometry techniques, including high-precision atomic accelerometers and gyroscopes, to measure forces acting on matter. An INS that used this technology would have unprecedented drift rates; however, many scientific and technical challenges remain. The vision is that PINS will operate with drift rates of less than 5 m/h hour and encompass a volume of less than 1 m^3. As this is entirely an inertial system, it will require no transmissions to or by the platform and will serve as a jam-proof, nonemanating INS with near-GPS accuracies for future military systems.

Deeply Integrated Guidance and Navigation Unit[14]

The goal of the deeply integrated guidance and navigation unit (DIGNU) program is the design, development, and implementation of automated manufacturing technologies for low-cost, high-accuracy, high-g, survivable microelectromechanical systems inertial measurement units (IMUs). ManTech funding will establish automated manufacturing technologies that will allow production of IMUs that meet 90 percent of DOD munitions, missiles, and—potentially—unmanned air system (UAS) needs. Phases 1 and 2 are complete, with both the IMU and DIGNU completing the government verification test. Phase 3 IMU was expected to have entered design verification testing in the first quarter of 2008.

Boeing iGPS System

Engineers from Boeing have invented and received patents (Cohen et al., 2007, 2005) on a series of techniques to assist GPS operation via integration with the existing Iridium low-earth orbit satellite system.[15] Boeing recently received a 3-year, $153.5 million cost-plus-fixed-fee contract from the Naval Research Laboratory to continue development of this system. The program is developing techniques that enable faster acquisition (time to first fix (TTFF)) of GPS satellite signals in adverse operating environments, including those with RF interference or urban settings. The high-integrity GPS team includes Boeing Advanced Systems and Phantom Works, Iridium LLC, Rockwell Collins, Coherent Navigation, and experts from academia.

Hostile Force Tagging, Tracking, and Locating

Systems for hostile force tagging, tracking, and location (TTL) are covered in detail by a separate NRC study.[16] Any future URFS will need to be able to interrogate and exfiltrate data and information from such systems and thus should be designed with compatible operating frequency, bandwidth, data rate, and range.

Fielded

ROVER

As seen in Figure 2, the remote observable video-enhanced receiver (ROVER) III, manufactured by L-3 Communications, Communication Systems-West, is a portable receive-only terminal that displays sensor data from multiple airborne platforms. It is interoperable with Ku-band digital (14.4-15.35 GHz), C-band digital (5.25-5.85 GHz), C-band analog (4.40-5.85 GHz),

---

[14]For more information on DIGNU, see http://www.armymantech.com/pg12.pdf. Last accessed September 29, 2008.

[15]For more information, see http://www.insidegnss.com/node/745.

[16]For more information on the NRC report Sensing and Supporting Communications Capabilities for Special Operations Forces: Abbreviated Version, see http://www8.nationalacademies.org/cp/projectview.aspx?key=48916. Last accessed September 8, 2009.

L-band analog (1.71-1.85 GHz), Predator, Shadow, Dragon Eye, Litening Pod, and other joint and coalition assets. The receiver weighs 10.25 lb and measures 3.8 in. x 5.5 in. x 15.5 in. (with battery). A single BA 5590 battery provides 10-12 hours of operation.

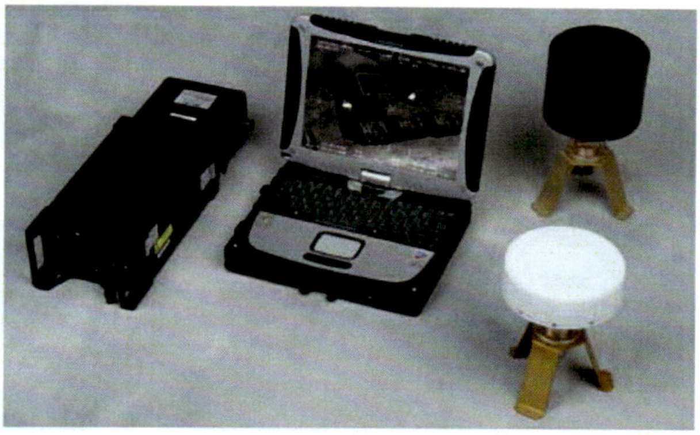

FIGURE 2 ROVER III. SOURCE: L-3 Communications (http://www.l-3com.com/products-services/docoutput.aspx?id=1259).

As seen in Figure 3, the ROVER 5 handheld, also manufactured by L-3 Communications, Communication Systems-West, is a portable transceiver device that provides sensor-to-shooter connectivity with the highest levels of collaboration. It is interoperable with Ku-band (14.4-15.35 GHz, 1.0 MHz steps), C-band (4.40-4.950 GHz, 1.0 MHz steps, and 5.25-5.85 GHz, 1.0 MHz steps), S-band (2.2-2.5 GHz, 0.5 MHz steps), L-band (1.71-1.85 GHz, 0.5 MHz steps), and UHF (400-470 MHz) among others. The receiver weighs 3.5 lb and measures 9.5 x 5.6 x 2.25 in. (with antenna). A lithium-polymer battery provides 2.5-3 hours of operation.

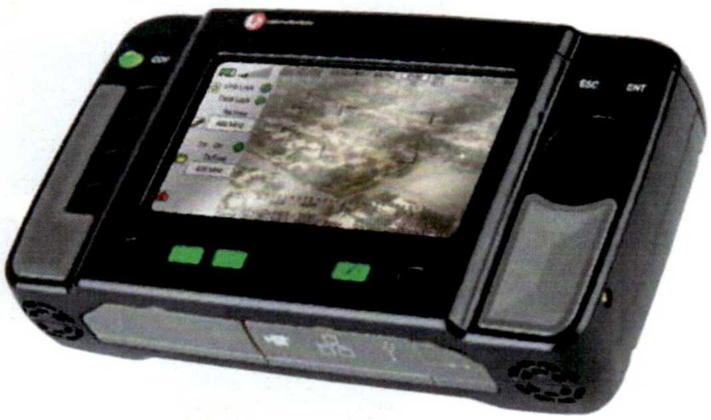

FIGURE 3 ROVER 5 handheld. SOURCE: L-3 Communications (http://www.l-3com.com/products-services/docoutput.aspx?id=1257).

Emerging Technology

DARPA Digital RF Tags

The DARPA digital RF tags (DRaFT) program, which ended in 2004, developed passive and active S- and X-band RF tags that are interrogated by existing airborne synthetic aperture radars (SARs) and ground moving target indicator (GMTI) radars such that location information is embedded in the radar return to provide information on the position of tagged vehicles. While primarily designed for preventing incidents of fratricide between close air support and friendly ground vehicles, DRaFT technology could also be used for hostile force TTL.

DARPA Dynamic Optical Tags

The dynamic optical tags program seeks to create new tagging, tracking, and location capabilities for U.S. forces. This program will develop optical tagging and interrogation technologies for small, environmentally robust, retro-reflector-based tags that can be read by both handheld and airborne sensors at significant distances. These tags can be used for unique, non-RF identification of items of interest or for monitoring tactical areas for disturbance by personnel and vehicles. The identification tags also will be capable of providing persistent two-way communications for both tactical and logistical operations.[17]

Signals Intelligence

Signals intelligence (SIGINT) is the capability to detect, identify, and geolocate enemy RF signals.

Fielded

AN/PRD-13v2

As seen in Figure 4, the AN/PRD-13v2, manufactured by L-3 Communications, Linkabit, is a manportable SIGINT system incorporating sophisticated RF intercept and direction finding processing capabilities into a low-power, lightweight, ruggedized, reliable system that satisfies the most demanding tactical applications and mission requirements. The system has a frequency range of 2-2000 MHz operating in the HF, VHF, and UHF bands. The transmitter has a power output of 9.5 W and weighs 43 lb, including an MB-5700 NiCd battery and all field accessories.

FIGURE 4 AN/PRD-13v2. SOURCE: L-3 Communications.

Available

DRT13013C

The DRT1301C, manufactured by Digital Receiver Technology, Inc., is a portable, ruggedized radio designed for operations in tactical and/or harsh environments. It provides a miniature yet powerful surveillance capability. The radio has a frequency range of 20-3000 MHz and operates against a variety of analog and digital wireless standards. The transmitter has a

[17]http://www.darpa.mil/STO/smallunitops/dots.html.

power output range of <1 W (standby) to 75 W (48 channels, 3 tuners); it weighs 10.5 lb and measures 3 in. (H) by 8.5 in. (W) by 11.2 in. (D).[18]

DRT13013C+

The DRT1301C+, manufactured by Digital Receiver Technology, Inc., is a portable, ruggedized manpackable radio offering the same capabilities as the DRT1401C but also has environmentally protected fans for additional cooling capability, permitting in-the-rucksack operation. The radio has a frequency range of 20-3000 MHz and operates against a variety of analog and digital wireless standards. The transmitter has a power output range of <1 W (standby) to 78 W (48 channels, 3 tuners); it weighs 12 lb and measures 3 in. (H) by 7.9 in. (W) by 13.32 in. (D).

Joint Threat Warning System Ground SIGINT Kit (PRD-14)

The Joint Threat Warning System (JTWS) ground SIGINT kit (PRD-14) is a manpackable SIGINT system developed for SOCOM by the SPAWAR Systems Center at San Diego. The JTWS, which completed Milestone C testing in 2004, consists of multiple receivers, tuners, and signal processing units and an operator station. It weighs ~45 lb (not including batteries) and consumes ~100 W of power.

Emerging Technology

DARPA Wolf Pack[19]

The DARPA WolfPack system, developed by BAE Systems, Inc. (Nashua, New Hampshire), is a complete end-to-end system consisting of remote sensors, advanced detection, tracking, and jamming algorithms, and a controller workstation capable of integration into a larger C4I system. The system emphasizes an air-deployable, ground-based, close-proximity, distributed, networked architecture to obtain RF spectrum dominance. The WolfPack program is developing new electronic warfare technologies that can hold enemy emitters (communications and radar) at risk throughout the tactical battle space while avoiding disruption of friendly military and protected commercial radio communications. The WolfPack system covers 30 MHz to 20 GHz in the RF spectrum.

DARPA BOSS[20]

The goal of the DARPA brood of spectrum supremacy (BOSS) program is to provide a RF-spectrum analogue to night vision capabilities for the tactical warfighter, with a particular focus on RF-rich urban operations. The program is intended to apply collaborative processing capabilities for software-defined radios to specific military applications. BOSS Phase I activities will be focused on modeling and simulation, resulting in hardware-independent, executable specifications of waveforms in a MATLAB format. Phase II is focused on implementing a prototype demonstration capability for an RF platform, with the implementation accompanied by hardware-independent, executable specifications of the waveforms. Phase III will focus on

---

[18]All dimensions are given in inches, and height, width, and depth are expressed as H, W, and D.

[19]For more information on DARPA's WolfPack program, see http://www.darpa.mil/STO/strategic/wolfpack.html. Last accessed September 29, 2008.

[20]For more information on DARPA's BOSS program see http://www.darpa.mil/ipto/Programs/boss/boss.asp. Last accessed September 29, 2008.

software communication architecture (SCA)-compliant waveforms suitable for implementation on a tactical software radio system.

Communications

Communications systems fielded by SOF come in manpackable and handheld packages. SOF radios have transitioned over the years from single-purpose (voice or data), single-band (HF, VHF, or UHF), single-waveform radios to multipurpose (voice and data), multiband RF systems capable of running multiple legacy waveforms or vendor-proprietary waveforms possessing proto-MANET capability. All of these systems have removable antennas.

Fielded

PRC-117F/C

As seen in Figure 5, the AN/PRC-117F/C, manufactured by Harris Corporation, is a multiband voice radio for a variety of military operations and has a frequency range of 30-512 MHz operating in the VHF-low, VHF-high, and UHF bands. The transmitter has a power output of 1-20 W; dimensions are 3.2 in. (H) by 10.5 in. (W) by 9.6 in. (D).The device weighs 12 lb and has a nominal power of 12 W. COMSEC interoperability includes TS KG84C, KY57, and ANDVT/KYV. Waveform interoperability includes AM/FM/PSK/CPM, SINCGARS ECCM, HAVEQUICK I/II, UHF DAMA/IW-HPW, and CTSS Tones. The radio has a removable keypad.

FIGURE 5 AN/PRC-117F/C. SOURCE: U.S. Air Force.

PRC-150

As seen in Figure 6, the AN/PRC-150, also manufactured by Harris Corporation, is a manpackable, tactical HF and VHF radio with a frequency range of 1.6-60 MHz. The transmitter has power outputs of 1 W, 5 W, and 20 W; dimensions (with battery case) are 10.5 in. (W) by 3.5 in. (H) by 13.2 in. (D). COMSEC interoperability includes ANDVT/KY-99, ANDVT/KY-100, KG-84C, KY-57 Vinson, and Citadel (export). Waveform interoperability includes 188-110B (modem at 75-9600 bps), Wideband FSK 16 kbps, 188-141B ALE, MELP/CVSD, and FED STD 1052 ARQ.

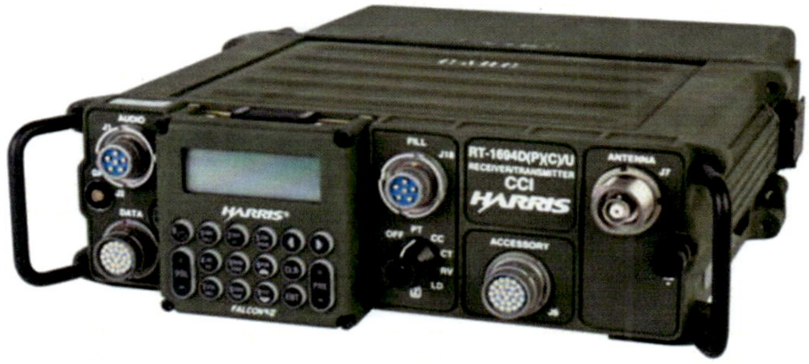

FIGURE 6 AN/PRC-150. SOURCE: Harris Corporation (http://www.rfcomm.harris.com/products/tactical-radio-communications/HB-AN-PRC-150C.pdf).

PSC-5C/D MBMMR

As seen in Figure 7, the AN/PSC-5C MBMMR, manufactured by Raytheon, is a lightweight, multiband/multimission terminal supporting critical tactical communications with a frequency range of 30-420 MHz operating in VHF and UHF. The transmitter has power outputs of 1 W, 5 W, and 20 W; dimensions are 3 in. (H) by 8.5 in. (W) by 10 in. (D). It weighs 10 lb and has a nominal power output of 12 W. COMSEC interoperability includes TS KG84C, KY57, and ANDVT/KYV-5. Waveform interoperability includes LPC-10 (ANDVT), MELP, SINCGARS, and HAVEQUICK I/II.

FIGURE 7 AN/PSC-5C MBMMR. SOURCE: Raytheon Company.

PRC-148

As seen in Figure 8, the AN/PRC-148 (the maritime and urban variants are known as V3 and V4, respectively), manufactured by Thales-TCI, is a widely fielded, handheld, multiband, tactical software-defined radio (SDR) with a frequency range of 30-512 MHz. The transmitter has a power output of 100 mW to 5 W; it weighs 30.6 oz and has a nominal power output of 3-5 W. COMSEC interoperability includes TS KG84C, KY57, and ANDVT/KYV-5. Waveform interoperability includes AM/FM/PSK/CPM, SINCGARS ECCM, HAVEQUICK I/II, UHF DAMA/IW, and CTSS Tones. The AN/PRC-148 is a Joint Tactical Radio System (JTRS)-approved SCA radio under the consolidated, interim, single-channel, handheld radio (CISCHR) contract with the JTRS Joint Program Executive Office.

FIGURE 8 AN/PRC-148. SOURCE: Thales Communications, Inc. (http://www.thalescommunications.com).

PRC-152

As seen in Figure 9, the FALCON III AN/PRC-152, manufactured by Harris Corporation, is a single-channel, multiband handheld radio with a frequency range of 30-512 MHz and adjustable transmit output power from 250 mW to 5 W. The AN/PRC-152 weighs 2.6 lb and has dimensions of 2.9 in. (W) by 9.6 in. (H) by 2.5 in. (D) (with battery). Interoperability includes SINCGARS, HAVEQUICK II, VHF/UHF AM and FM, and MIL-STD-188-181B. As with the AN/PRC-148, the AN/PRC-152 is also a JTRS-approved SCA radio under the CISCHR contract with the JTRS JPEO.

FIGURE 9 FALCON III AN/PRC-152. SOURCE: Harris Corporation.

PRC-343

As seen in Figure 10, the Selex PRC-343, also known as the Personal Role Radio (PRR), uses 2.4-GHz spread-spectrum technology to provide push-to-talk (PTT) voice functionality. Some key features of the PRR are 50 mW of transmit power using direct sequence spread spectrum (DSSS) modulation, a typical operating range of 500 m in open terrain, and through three floors of a building or through five houses in urban environments. Sixteen channels are available directly to the user; it operates on two AA batteries for more than 24 hr (1:7:16

TX/RX/standby) and is independent of any infrastructure. The PRR was originally developed for use with the British Army as the first phase of the Bowman project.[21]

FIGURE 10 Selex PRC-343. SOURCE: Selex.

Iridium

As seen in Figure 11, the 9505A, manufactured by Motorola, is a small, lightweight satellite phone resistant to water, dust, shock, and environmental variables and is ideal for remote areas and rugged conditions. The phone offers up to 30 hours of standby time and 3.2 hours of talk time. It weighs only 13.2 oz and has dimensions of 6.2 in. (H) by 2.4 in. (W) by 2.3 in. (D). Iridium runs on a constellation of LEO satellites operated by Boeing. Iridium systems with the CONDOR appliqué have the ability to do Type 1 voice encryption.

FIGURE 11 Iridium 9505A satellite phone. SOURCE: Iridium.

PRC-112

The AN/PRC-112 survival radio, manufactured by General Dynamics, provides search and rescue (SAR) personnel with the ability to perform combat search and rescue (CSAR) missions to save downed aircrew. The PRC-112 is used extensively for personnel recovery as well as for various SOF applications. The PRC-112 is a software-defined radio (SDR) capable of running the following waveforms: AM voice, DME, LOS HOOK data, over-the-horizon (OTH) UHF SATCOM data, 406 SARSAT, 121.5 and 243 MHz swept-tone beacon and other special waveforms. It has an output power of 1-5 W (UHF) and 0.1 W (VHF). It weighs 28 oz (with battery) and measures 7.69 in. (H) by 3 in. (W) by 1.5 in. (D).

---

[21]For more information on PRC-343, see http://www.janes.com/articles/Janes-Military-Communications/H4855-AN-PRC-343-Personal-Role-Radio-PRR-United-Kingdom.html. Last accessed December 18, 2008.

Available

The Falcon III AN/PRC-117G(V)1(C)

As seen in Figure 12, the Falcon III AN/PRC-117G(V)1(C) is manpackable. Manufactured by Harris Corporation, it is a tactical radio with a frequency range of 30 MHz to 2 GHz and has adjustable transmit output power of 10 W VHF and 20 W UHF. The AN/PRC-117 weighs 10.9 lb, has dimensions of 7.4 in. (W) by 3.7 in. (H) by 13.5 in. (D) (with battery), and is submersible to 1 m. Waveforms include SINCGARS, HAVEQUICK II, VHF/UHF AM and FM, high-performance waveform, MIL-STD-188-181B SATCOM, and the Harris Adaptive Networking Wideband Waveform (ANW2).

FIGURE 12 Falcon III AN/PRC-117G(V)1(C) is manpackable. SOURCE: Harris Corporation.

SINCGARS RT-1523

As seen in Figure 13, the SINCGARS RT-1523, manufactured by ITT Industries, is designed to provide network data services in both mounted and dismounted configurations to allow access to the Tactical Internet. The radio has a frequency range of 30-88 MHz with transmitting power options of 1 mW, 100 mW, and 5 W dismounted and 50 W mounted RF power amplifier (RFPA). With embedded battery, the radio weighs 7.7 lb; dimensions are 3.4 in. (H) by 5.3 in. (W) by 10.15 in. (D).

FIGURE 13 SINCGARS RT-1523. SOURCE: ITT Industries.

SINCGARS SIP/RT-1523E

As seen in Figure 14, the SINCGARS SIP (ASIP), manufactured by ITT Industries, is the primary combat net radio for the U.S. Army, designated primarily for voice C2 for infantry, armor, and artillery units. The radio incorporates all the features of previous radio systems and enhancements to reduce its weight and size for the dismounted soldier and optimize its performance in the Tactical Internet. This is mainly due to the internal redesign of the radio and to taking advantage of software-based digital signal processing architecture. It has a frequency range of 30-88 MHz VHF-FM; the transmitter has a nominal power output of 4-5 W. With battery, handset, and antenna, the total manpack weight is less than 9 lb. Dimensions are 3.4 in.

(H) by 5.3 in. (W) by 10.15 in. (D). It has integrated COMSEC and data rate adapter with embedded internet controller and SAASM GPS options.

FIGURE 14 SINCGARS SIP (ASIP). SOURCE: ITT Industries.

CSEL (AN/PRQ-7)

The combat survivor evader locator (CSEL) is the DOD Program of Record for Joint Search and Rescue and is in full production by prime contractor Boeing. This fully qualified, next-generation survival radio system comprises OTH relays, ground, and user equipment segments for the joint services. CSEL minimizes the search aspect of a rescue mission by providing recovery forces with precise geopositioning information and secure OTH and LOS two-way data communications capabilities. The CSEL is a multifunctional handheld radio that gives the warfighter the ability to securely communicate position and text messages through the CSEL UHF SATCOM network. The CSEL radio has a programmable and upgradable software, a SAASM-based GPS receiver for precise navigation, and receives OTH waypoints and text messages. All OTH transmissions are acknowledged and it has National Security Agency (NSA)-certified encryption and decryption of OTH and LOS messages.

EPLRS MicroLight

As seen in Figure 15, the EPLRS MicroLight-DH500, manufactured by Raytheon, is a fully networked radio for RF-challenged environments. It integrates voice, data, and video transmission into a single low-profile radio. The MicroLight has a frequency range of 225-2000 MHz, a power output of 100 mW to 4 W, and a weight of 27 oz. The dimensions are 7.65 in. (H) by 2.6 in. (W) by 1.6 in. (D) (including battery). MicroLight operates in the 420-450 UHF band utilizing waveforms for interoperability with various data networks, including JTRS, EPLRS UAF, and SRW as well as the Air Force's SADL.

FIGURE 15 EPLRS MicroLight-DH500. SOURCE: Raytheon Company.

Soldier Radio

As seen in Figure 16, the Soldier Radio, manufactured by ITT Industries, is an SDR designed for PTT voice and data communications. Soldier Radio supports narrowband and wideband waveforms with a frequency range of 30-88 MHz (VHF), 225-970 MHz (UHF), and 1650-1850 MHz (L-band). Programmable transmit power is 1 mW to 5 W in ~6 dB steps. Weight is less than 2.2 lb; dimensions are 5.99 in. (H) by 3.75 in. (W) by 1.9 in. (D).

FIGURE 16 Soldier Radio. SOURCE: ITT Industries.

SpearNet Team Member Radio

As seen in Figure S-17, the SpearNet Team Member Radio, manufactured by ITT Industries, is optimized for digital, network-centric communication, providing a seamless, self-healing ad hoc networking and multihop routing capability. Standard interfaces include Ethernet, USB, RS-232, and Bluetooth, and it has embedded GPS. It has a 1.2 GHz operating band with transmit power up to 26 dbm. The radio weighs 1.5 lb and has dimensions of 7.72 in. (H) by 2.99 in. (W) by 1.20-1.87 in. (D).

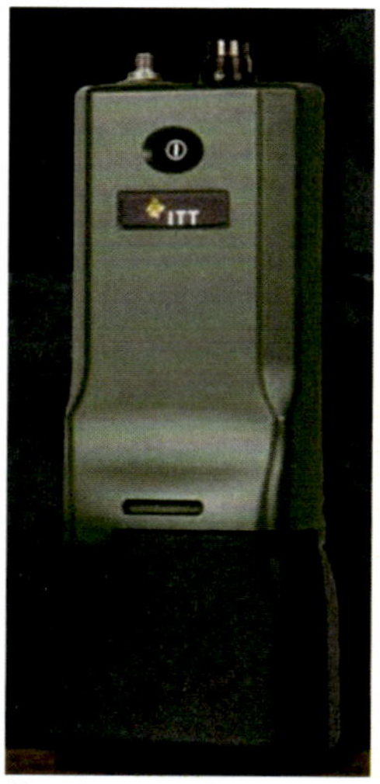

FIGURE 17 SpearNet Team Member Radio. SOURCE: ITT Industries.

Wearable Soldier Radio Terminal

As seen in Figure 18, the Wearable Soldier Radio Terminal (WSRT), manufactured by ITT Industries, provides communication capability for dismounted soldiers, linking soldiers to each other and bridging the dismounted gap. WSRT is an SDR capable of supporting both narrowband and wideband waveforms with Type II COMSEC and interfaces with Ethernet, USB, and headset. Weighing only 1.2 lb, the system operates in the UHF band with transmit power up to 5 W.

FIGURE 18 Wearable Soldier Radio Terminal. SOURCE: ITT Industries.

Cobham Eagle

As seen in Figure 19, the Cobham Eagle is a short-range infantry assault radio designed and developed by Cobham Defense Communications to provide a wireless extension for the ROVIS, AN/VIC-3, and LV2 intercom systems. The equipment is a fourth-generation, full-duplex networked radio specifically designed to provide the soldier with a low-cost solution for voice and data requirements in the urban and rural environments of the modern battlefield. Eagle is designed to enhance C2 at the section and squad levels. In comparison to earlier first-generation simplex systems, Eagle allows access for up to six full-duplex users simultaneously, utilizing advanced VOX to ease the burden of communications under stress and avoid jamming the system in situations when all users are most likely to want access to the network at the same time. In addition to the voice functions of Eagle, simultaneous transmit and reception of data are also incorporated into the design of the radio, allowing up to 128 kbps of data while maintaining two full-duplex channels for audio plus a permanent emergency override capability. Eagle provides extended area and range coverage via automatic network extension. Eagle is an SDR operating in the license-free, 2.4-GHz ISM band. The use of a DSSS frequency-hopping waveform provides LPI and LPD and the ability to coexist with other networks operating in the same band without degradation of voice and data communications capabilities.

FIGURE 19 Cobham Eagle. SOURCE: http://www.cobhamdes.com/img/content/products-eagle-black-pouch.jpg.

Emerging Technology

JTRS HMS

As seen in Figure 20, the JTRS HMS manpackable radio is a two-channel SDR capable of network-centric connectivity and legacy interoperability, supporting advanced and current-force waveforms. The JTRS HMS has a frequency range of 2 MHz to 2.5 GHz and a maximum power output of 20 W. Weight is 14.5 lb (with battery), size is 400 cu. in., and it has fully programmable COMSEC and TRANSEC interoperability. Waveforms include SRW, MUOS, SINCGARS, EPLRS, SATCOM, and HF SSB w/ALE. General Dynamics is the prime contractor.

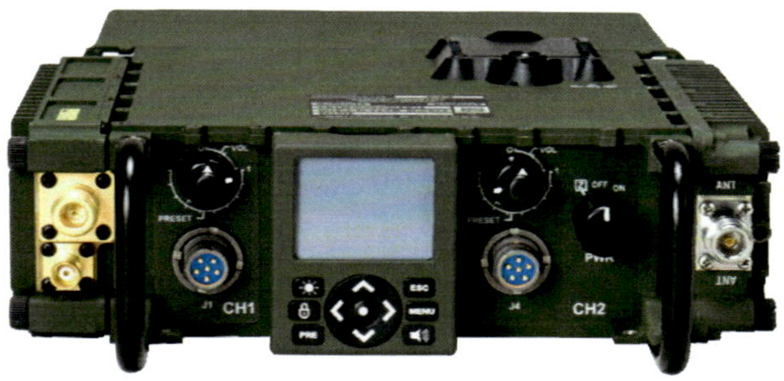

FIGURE 20 General Dynamics JTRS HMS Manpack. SOURCE: General Dynamics (http://www.gdc4s.com/jtrshms).

DARPA Wireless Network after Next

The Wireless Network after Next (WNaN) program goal is to develop and demonstrate technologies and system concepts enabling densely deployed networks in which distributed and adaptive network operations compensate for the limitations of the physical layer of the low-cost wireless nodes that comprise these networks. WNaN networks will manage node configurations and the topology of the network to reduce the demands on the physical and link layers of the nodes. The technology created by the WNaN effort will provide reliable and widely available battlefield communications at low system cost ($500 per unit in lots of 100,000). The WNaN program will develop a prototype handheld wireless node that can be used to form high-density ad hoc networks and gateways to the global information grid. It will develop robust networking architecture(s) that will exploit high-density node configurations from related DARPA programs. This program will culminate in a large-scale network demonstration using inexpensive multichannel nodes. Cobham (formerly M/A Com) is the prime contractor for the radio hardware, and BBN Technologies, Inc., is the prime contractor for the network software.[22]

QNT

The Quint Networking Technology (QNT) is a modular network data link program focused on providing a multiband modular capability to close the seams between five nodes: aircraft, unmanned combat air vehicles (UCAVs), weapons, tactical UASs, and dismounted ground forces. The specific intended QNT hardware users are weapons, air control forces on the ground (dismounted), and tactical UASs. The desired QNT data link functional capability includes the ability to transmit target coordinates to a weapon in flight from either an aircraft or a dismounted ground unit; to disseminate sensor data; to alter the missions of dismounted ground units, unmanned airborne platforms, and weapons; to allow autonomous bomb impact assessment or

[22] For more information on WNaN, visit http://www.darpa.mil/sto/strategic/wireless.html.

bomb hit indications; and to enable offboard sensor data dissemination and control supporting cooperative engagement. Rockwell Collins is the prime contractor for QNT.[23]

Commercial

There are a plethora of commercial communications systems (Table 1) that could be used by SOF in certain circumstances as permitted by the CONOPS and depending on security, covertness, and the availability of infrastructure. The committee covers only the Inmarsat BGAN system currently used by SOF but provides URLs for commercial commodity communications systems in the table.

TABLE 1 Commercial Communications Systems

Communications System	Information Available at http://en.wikipedia.org/wiki/___
Cellular GSM voice and data	GSM
Cellular CDMA voice and data	CDMA
802.11 (WiFi) derivatives	WiFi
802.16 (WiMax) derivatives	WiMax
802.15.4 (ZigBee)	Zigbee
BlueTooth	Bluetooth
APCO-25	APCO-25
TETRA	TETRA

Inmarsat Broadband Global Area Network (BGAN)

Inmarsat BGAN offers data rates of up to hundreds of kilobits per second in a manpackable (~2 kg) satellite terminal supporting voice and data services and is available globally with the exception of the extreme polar regions. Inmarsat BGAN is used by, for example, commercial news-gathering services, executives on private jets, oil and gas companies, and others that need moderate-data-rate Internet connectivity in austere environments without telecommunications infrastructure.

SYNOPSIS OF FINDINGS AND RECOMMENDATIONS

SOCOM should expeditiously acquire multifunctional handheld and, if necessary, manpackable systems to replace multiple, separate single-function systems. Recognizing that power and energy needs are mission-critical, SOCOM's acquisition process should give the power-energy issue a very high priority. It should demand greater efficiencies by aggressively pursuing energy conservation and management in device design and by exploiting emerging battery and power-source technologies.

The committee's examination of the current state of the art revealed a plethora of radio equipment: legacy radios intermingled with newer RF systems and some systems nearly identical in function. After reviewing a wide range of research and technology development efforts to assess what capabilities could be provided in the years ahead, the committee found that existing and emerging technologies could support many capabilities and attributes relevant to future universal RF systems. Accordingly, SOCOM should establish a structured program for a universal RF system to address far-term research and technology issues. That program should incorporate relevant efforts of the research and technology development community at large.

[23]For more information on QNT, visit http://www.darpa.mil/ipto/programs/qnt/qnt.asp.

Looking to deployability, SOCOM should evaluate (1) existing software-defined radios in terms of near-term upgrades, long-term growth, and affordability and (2) the costs and benefits of upgrading existing platform(s). Additionally, SOCOM should conduct an expeditious, rough-order-of-magnitude analysis of cost as well as the operational and SWaP benefits of eliminating the future procurement of some manpackable systems in favor of handheld-only systems.

SOCOM should adopt a modular approach. The long-term goal should be incorporation of ASIC system-on-a-chip (SoC) processors for high-speed signal processing (a goal that will require strong technical oversight), and SOCOM should look to a universal ASIC chip capable of synthesizing most, if not all, likely waveforms.

With respect to power supplies, SOCOM should leverage Army developments in advanced lithium batteries, lithium–rechargeable zinc/air hybrids, energy recovery from partially discharged batteries, and low-observable, lightweight, flexible photovoltaic systems for battery charging. For the far term, SOCOM should develop mature hybrid power systems, standard batteries with common interfaces, and smart battery chargers.

The following section contains publicly releasable findings and recommendations. The full findings and recommendations are provided in a separate full report.

SELECTED FINDINGS AND RECOMMENDATIONS

Finding 2-1. Many of the capabilities needed by SOF are unique and much more demanding than those of conventional forces, particularly with respect to attributes related to covert operations. Security; size, weight, and power; frequency agility; and the incorporation of multipurpose functionality are essential to effective use by small teams operating in uncontrolled and hostile areas.

Recommendation 2-1. SOCOM should (1) expeditiously acquire multifunctional handheld and, if necessary, manpackable systems to replace the multiple separate single-function systems now carried and (2) aggressively develop further consolidated multifunctional platforms with currently available technology.

Finding 2-2. Greater and more efficient energy utilization is needed and, if feasible, could be realized by sharing energy between devices to increase mission effectiveness.

Recommendation 2-2. Standard and robust power connectors/adapters and smart chargers should be developed so that batteries and other power sources can be shared between devices.

Finding 2-3. Power and energy requirements are mission critical, needing increased capability and reduced operator weight and volume burden.

Recommendation 2-3. SOCOM's acquisition process should make the power-energy issue a very high priority and demand greater efficiencies by aggressively pursuing energy conservation and management in device design through hardware and software improvements and the exploitation of emerging battery and power source technologies. This should be carried out in parallel and in the context of likely SOCOM mission scenarios. SOCOM should heed the relevant recommendations of two previous NRC reports as they apply to its unique mission needs (see Appendix C).

Finding 2-4. The SOF URFS requirement is so unique and demanding that conventional research and development is very unlikely to meet it.

Recommendation 2-4. SOCOM should continue to engage leading universities, the national and service laboratories, and industry in grand challenges to meet SOF-unique requirements, e.g., small, efficient, agile, multiple-band antennas and enhanced power sources.

Finding 2-5. The cost of even a specially developed URF system is very small compared to the overall mission costs and returns from critical SOF mission accomplishment.

Recommendation 2-5. SOCOM and the DOD should expeditiously procure and field the specialized URFS equipment necessary for SOF mission success.

Finding 3-1. The primary differences between handheld and manpackable radio systems are the transmit power amplifier (PA) power and the battery.

Finding 3-2. In the past, the need for mission flexibility drove SOCOM to unique solutions, resulting in a large inventory. As a result, some manpackable and handheld systems are nearly identical in function yet they coexist in the current inventory. That raises the issue of how SOCOM might be able to reduce the number of such systems to streamline training and logistics while maintaining competition and industrial knowledge base.

Finding 3-3. Most current RF systems use omnidirectional, narrowband (e.g., 25 kHz), single-polarization antenna systems, which are not compatible with emerging wider band systems. A transition to the wideband systems necessary for adding new functionality would require new RF transmit and receive (LNAs, PAs, filters, ADCs, dual-polarization antennas, etc.) architectures.

Finding 3-4. Many of the existing state-of-the art RF systems of interest are SDRs that are not necessarily compatible with the DOD JTRS software communications architecture (SCA).

Finding 3-7. Warfighter load is exacerbated by the number and type of RF systems and batteries. Advanced batteries, energy conversion and/or hybrid technologies, and energy conservation measures have the potential to significantly decrease operator SWaP burdens. However, the introduction of energy conversion technologies must be traded against operation with batteries alone with respect to size, weight, logistics, operational factors, and thermal and acoustic signatures.

Recommendation 4-1. SOCOM should establish a structured URFS program to address the far-term technology research issues. The program should include the formal feedback of lessons learned from the operational use of near- and medium term capabilities identified above. The program should also incorporate relevant efforts from the research and technology development community at large—for example, the military service laboratories, the Defense Advanced Research Projects Agency (DARPA), national laboratories, universities, industry, and appropriate foreign sources.

Appendixes

Appendix A

Biographical Sketches of Committee Members

Lawrence Delaney, Chair, is retired executive vice president of operations and president of the Advanced Systems Development Sector of Titan Corporation. Previously, he held positions with Areté Associates, Inc.; Delaney Group, Inc.; BDM Europe; and the Environmental and Management Systems Group at IABG. He was also the Acting Secretary of the Air Force and served as the Assistant Secretary of the Air Force for acquisition, as well as the Air Force's service acquisition executive, responsible for all Air Force research, development, and acquisition activities. He provided direction, guidance, and supervision of all matters pertaining to the formulation, review, approval, and execution of acquisition plans, policies, and programs. Dr. Delaney has more than 41 years of international experience in high technology program acquisition, management, and engineering, focusing on space and missile systems, information systems, propulsion systems, and environmental technology. He is vice chair of the Army Science Board and served as chair of the NRC Air Force Studies Board and as a member of the NRC Board on Army Science and Technology and the NRC Standing Committee on Research, Development, and Acquisition Options for Special Operations Command (SOCOM).

Brian Agee is currently the president of B3 Advanced Communication Systems. His engineering activities include development of engineering tools for design, analysis, and visualization of collaborative communication and reconnaissance networks, including terrestrial, airborne, and satellite communication systems and networks; development of robust emitter detection and geolocation methods for collaborative reconnaissance networks; development of techniques and systems for detection, demodulation, and geolocation of WLAN devices; and development, analysis, and simulation of transceivers for secure, adaptive MIMO mesh networks. Dr. Agee also currently holds the position of adjunct research (full) professor at the Virginia Polytechnic Institute, where he is performing research and has guided graduate students in development of collaborative communication and signal analysis networks and in development of algorithms and techniques for characterization (detection, identification, and geolocation) and management of structured and unstructured interference in conventional and cooperative 802.11 enterprise networks. From 2001 to 2004, Dr. Agee was cofounder, president, and chief technical officer of Protean Radio Networks, a developer of spatially adaptive transceivers and mesh networking technology for wireless communication systems and networks. From 1998 to 2001, Dr. Agee was self-employed as a consulting engineer and performed independent research and development of spatially adaptive transceivers and mesh networking technology for wireless communication systems and networks and for geolocation of cellular telecommunication signals in urban multipath environments. From 1990 to 1998, Dr. Agee was director of Engineering Studies (1990-1996) and director of Advanced Concept Development (1996-1998) at Radix Technologies, Inc. (now a division of Argon ST), which is a developer of systems, equipment, and technology for commercial and U.S. government-funded communication, collection, and analysis systems. As part of his responsibilities at Radix, he developed or led development of the core system concepts and algorithms for all of Radix's major signal collection and analysis projects and developed or led development of the core system concepts and algorithms for all of Radix's telecommunications projects, including the first point-to-multipoint cellular air interface employing TDD-OFDM modulation formats and fully adaptive/reciprocal antenna arrays as part of AT&T's project Angel. He is a senior member of the IEEE. Dr. Agee holds a Ph.D. in electrical engineering from the University of California at Davis.

Mark Buckner is director of the Cognitive Radio Program at Oak Ridge National Laboratory. In this post he is leading a team of innovative scientists and engineers to integrate software radio, sensors, and computational intelligence capabilities to realize the art of the possible in cognitive computing and communications to address both government and commercial problems in a manner that enhances U.S. national security. He is currently developing a crosscutting initiative in bioinspired computing, communication, and signal processing, is principle investigator for a number of research initiatives for the intelligence community, and is working on the application of the global optimizer and other soft-computing techniques to problems in the software-defined and cognitive radio domains. Since joining the Oak Ridge National Laboratory in 1987, he has also championed innovations in the design and development of nuclear/radiation sensor systems and smart networkable transducers; system-level modeling, simulation, and rapid prototyping of hardware and software communications platforms; soft-computing techniques for nonlinear engineering problems; mobile ad hoc wireless sensor systems; and sensor- and GPS-enabled RFID tagging, tracking, and locating systems for national security needs. He is the author or coauthor of over 40 conference and refereed papers related to sensors, communications, and RFID and is frequently invited to speak in seminars and workshops. He has served on the DOD RFID Technical Advisory Working Group, AIM Global RFID Experts Group, and ANSI/INCITS T6 RFID Working Group and is a member of the Software Defined Radio Forum. Dr. Buckner holds a Ph.D. in nuclear engineering from the University of Tennessee.

R. Michael Buehrer joined Virginia Polytechnic Institute and State University from Bell Labs as an assistant professor with the Bradley Department of Electrical Engineering in 2001. He is currently an associate professor and is a member of Wireless @ Virginia Tech, a comprehensive research group focusing on wireless communications. His current research interests include dynamic spectrum sharing, cognitive radio, MIMO communications, intelligent antenna techniques, position location networks, ultrawideband, spread spectrum, interference avoidance, and propagation modeling. His work has been funded by the National Science Foundation, the Defense Advanced Research Projects Agency, Office of Naval Research, and several industrial sponsors. Dr. Buehrer has coauthored over 35 journal articles and 100 conference papers and holds 11 patents in the area of wireless communications. He is currently a senior member of the IEEE and an associate editor for IEEE Transactions on Wireless Communications and IEEE Transactions on Vehicular Technologies. Previously he served as an associate editor for IEEE Transactions on Signal Processing and IEEE Transactions on Education. In 2003 he was named Outstanding New Assistant Professor by the Virginia Tech College of Engineering.

John Cafarella is an independent consultant who has contributed to the development of the first single-chip SAASM GPS receiver, with responsibility for the overall system design, the digital signal processing, and ASIC architecture. Dr. Cafarella demonstrated the first engineered SAW acoustoelectric convolver in 1975 and subsequently pursued its application to spread-spectrum communications. Having extensive knowledge of devices, circuits, and signal processing, he established an activity to integrate emerging signal-processing components into system-level demonstrations. This included development of advanced spread-spectrum radios and processors for tactical radars. In 1984, Dr. Cafarella left Lincoln Laboratory to found MICRILOR, which specialized in applications of advanced signal-processing technologies. Dr. Cafarella's work at

MICRILOR included analysis and systems design for applications including spread-spectrum radio, underwater communications, tactical radars, and intelligence systems, as well as support of DARPA and other U.S. government agencies in development of advanced technologies. Dr. Cafarella directed the MICRILOR team in the development of the first 10-Mbps spread-spectrum wireless LAN technology and its application to commercial products for Clarion, Ltd., of Japan. He is coholder of the patent for the spread-spectrum signaling technique first used to achieve a 10-Mbps rate under FCC part 15.247. Dr. Cafarella is a fellow of IEEE and participated in IEEE deliberations from 1997 to 1998 in the adoption of new high-speed wireless LAN standards 802.11a and 802.11b. Dr. Cafarella serves on the Army Science Board and is a member of the study on Army LandWarNet communications. Dr. Cafarella holds an Sc.D. from the Massachusetts Institute of Technology.

Phil Dickinson is an independent consultant with broad experience and expertise in command, control, communications, intelligence, field experimentation, and operational test and evaluation. During his career as an Army employee, he held positions of increasing authority, including deputy director for battlefield systems integration, a combined TRADOC, at the U.S. Army Materiel Command organization; director Battlefield Exploitation Targeting and Analysis (BETA), a DARPA-initiated Joint Intelligence for NATO; technical director of the U.S. Army Operational Test and Evaluation Agency; and acting deputy assistant secretary of the Army for programs and plans. He was the director of army studies and analysis at E-Systems Center for Plans and Analysis. He is an active member of the Army Science Board, where he chaired or co-chaired nearly 10 studies related to Army C4ISR. Dr. Dickinson holds a Ph.D. in electrical engineering from the University of Florida.

James Freebersyser is the director for advanced systems at BBN Technologies, Inc., with expertise in communications systems including terrestrial and satellite communications, internetworking, and mobile, wireless (ad hoc) networks. He has held industry positions at Honeywell Aerospace Laboratories, GE AstroSpace, and GTE Government Systems and government positions in the Advanced Technology Office (ATO) of DARPA, the Office of Naval Research, and the Army Research Office. Dr. Freebersyser holds a Ph.D. in electrical engineering from North Carolina State University.

Rita Gonzales is the manager of the Mixed Signal ASIC/SoC Products Group at Sandia National Laboratories. Her department specializes in developing extremely high reliability digital and analog application-specific integrated circuits (ASICs) products, including focal plane array pixel readout electronics, secure processors, high-performance digital radio transmitters/receivers, and trusted microelectronics for the DOE weapons community, the DOD armed forces, the intelligence community, and commercial industry. Before managing the group, Ms. Gonzales worked as a member of the technical staff designing and leading design teams in realizing ASICs for the aforementioned applications. She also worked in the Sandia satellite community supporting hardware design for nonproliferation programs. Ms. Gonzales holds an M.S. in electrical engineering from Stanford University.

Andy Ivers is president of L-3 Communications, Linkabit Division, which was acquired by L-3 in July 2005. Mr. Ivers joined Linkabit in 2001 as senior vice president of engineering and took over the leadership of the division in 2004 as president. Mr. Ivers has over 20 years of experience in design engineering for a variety of communications systems, including HF, VHF, UHF tactical and strategic radios, LPI communication, and SATCOM communication, including work on the MILSTAR ground terminals and numerous nonprotected links. In addition to working for L-3, Mr. Ivers worked for Harris RF Communications Division as the director of tactical radio engineering, Raytheon Equipment Division in the Communications Systems Laboratory, Hughes

Aircraft, and Litton Systems. Mr. Ivers has an extensive systems engineering and software background. Earlier in his career, while employed at Raytheon, Mr. Ivers managed system engineering activities on numerous communications systems programs, including SHF SATCOM modem designs; L-band data link designs; LPI/LPD HF communications systems; and HF and W-band antifratricide system designs; he also participated in the initial SMART-T prototype design for the EHF SATCOM modem. Mr. Ivers graduated with an M.B.A. from the Simon School, University of Rochester, and has a B.S.E.E. and a B.S. in math from the University of California, Irvine, as well as numerous graduate-level courses in communications systems engineering from California State University Northridge.

Peter Kind is a research staff member at the Institute for Defense Analyses, where he leads and supports research, analyses, assessments, and projects in information technology, homeland security, national intelligence, command and control, software-intensive systems, telecommunications, modeling and simulation, collaboration, and logistics. LTG Kind has extensive command center experience at all military levels, including interface with interagency and coalition forces. He created and directed the White House Information Coordination Center as the government focal point for status and events during the Y2K millennium rollover. He also advised and led program development and operations of the White House Office of Homeland Security Coordination Center. As director of information systems for C4 of the U.S. Army, he directed plans, policy, programming, budgeting, and execution of information technology and communications services and support for the U.S. Army, federal agencies, and supported activities. In addition, as commander of the U.S. Army Signal Center, he directed the development of new equipment and signal doctrine. He is a graduate of the U.S. Army War College and holds an M.B.A. from Harvard Graduate School of Business Administration.

Gary Minden is professor of electrical and computer science at the University of Kansas. From 1971 to 1978, Dr. Minden was a research engineer at the University of Kansas Center for Research, Inc. During that period he worked on problems in the areas of image-processing systems, multiprocessor computer systems, and general systems theory. From 1978 to 1980 he was a vice president of CHILD, Inc., where he was a co-designer of the LIGHT-50 computer graphic terminal. In 1981, he joined the University of Kansas Department of Electrical Engineering as an assistant professor. From 1983 to 1989, he led the implementation of a new computer engineering degree program within the Electrical and Computer Engineering Department. In 1991, Dr. Minden completed a sabbatical at Digital's System Research Center, working on gigabit local area networks. He is a principal investigator on the MAGIC gigabit testbed and the Rapidly Deployable Radio Network (RDRN) at ITTC. From 1994 to 1996, he was on leave from DARPA's Information Technology Office, where he served as a program manager in the area of high-performance networking systems. While at DARPA, he formulated and initiated a new research program in active networking.

Joseph Mitola III is an internationally recognized expert on software-defined and cognitive radio systems and technologies at the Stevens Institute of Technology. In addition to having published the first paper on software radio architecture in 1991, he has taught short courses on software radio in the United States, Asia, and Europe. He was founding chair of the SDR Forum in 1996 and was first to receive the Forum's Achievement Award. In his 1999 Licentiate in Teleinformatics, he coined the term "cognitive radio" to refer to technologies integrating machine perception of vision and language with machine learning into software radio. Dr. Mitola published the first interdisciplinary graduate text on software radio, Software Radio Architecture, and the first graduate text on cognitive radios, Cognitive Radio Architecture. Prior to MITRE, Dr. Mitola was the chief scientist for electronic systems at E-Systems Melpar Division, culminating a career at E-Systems that had begun in 1976. He has also held positions of technical leadership with Harris Corporation, Advanced Decision Systems, and ITT Corporation. Dr. Mitola holds a doctorate in teleinformatics from the Royal Institute of Technology, Stockholm.

Robert Nowak is a private consultant. He was a program manager at DARPA and the Office of Naval Research and has directed and supported research in fundamental electrochemistry, fuel cells, batteries, capacitors, energy harvesting, fuel processing, thermal energy conversion, microengines, hydrogen storage, biofuel cells, sonoluminescence, and biomolecular motors. He received a Ph.D. in chemistry from the University of Cincinnati. He was a postdoctoral research associate at the University of North Carolina at Chapel Hill and was selected as NRC postdoctoral fellow at the Naval Research Laboratory, where he continued his research activities in polymer electrochemistry and chemically modified electrodes as a staff scientist and section head. Dr. Nowak received the Secretary of Defense Meritorious Civilian Service Award in 2002 for his efforts in developing portable power sources for the military. He was a member of the NRC Committee on Soldier/Power Energy Systems, the NRC Panel on Benefits of Fuel Cell R&D, the NRC Committee Review of the Research Program of the Freedom Car and Fuel Partnership, and the NRC Standing Committee on Research, Development, and Acquisition Options for Special Operations Command (SOCOM).

Glen Roussos is currently the vice president for Army ISR and SATCOM business in L-3 Communications, Communication Systems-West (CS-W). In this role he is responsible for Army and international programs. The programs in his area include the Phoenix (AN/TSC-156), communications for the Guardrail system and international SATCOM business. Prior to his current position, he was the director for SATCOM programs, where he oversaw Army and SOCOM SATCOM programs. He has also been a business development manager for Army programs, during which he worked on the Army's Future Combat Systems. Mr. Roussos joined CS-W after 22 years of federal service both as an Army officer and a government service employee. He retired from the Army in 2004. During his Army career, he served in command and operations positions in the Special Forces. As a Special Forces commander, he led missions during Operations JUST CAUSE and JOINT GUARDIAN in Bosnia. He also led deployments to Colombia and Venezuela. His involvement spanned the spectrum of Special Forces missions, which included Direct Action, Special Reconnaissance, and Foreign Internal Defense in Central and South America and Eastern Europe. Upon his retirement from the Army, he accepted a GS-15 position as a division chief in NORTHCOM and led a division that conducted force structure analysis for homeland defense. He has a bachelor's degree in industrial distribution, a master's in applied mathematics, and a master's of business administration.

Robert Tingley is presently manager of the RF and communications group at the Charles Stark Draper Laboratory. He has 20 years' experience in research, engineering, product development, and management of radio frequency, communication, and signal processing systems for a broad range of defense, intelligence, and commercial customers. As task leader for the ARINC network evaluation project, Dr. Tingley led a team that developed a detailed model and associated discrete event simulation of a commercial worldwide air-to-ground data communication network. Serving in the role of communications system task leader of the ONR program RSVP, he led the team that developed propagation models for RF propagation aboard naval destroyers, designed and manufactured ultra-low-power wireless sensor nodes, and invented and patented new techniques for first-responder tracking. As task leader for the DARPA tagging, tracking, and locating program he evaluated state-of-the-art technology and proposed and won funding for two follow-on development efforts: the radiant oscillator and GPS transponder tags. He was also task leader for the DARPA program multipath navigation, which sought to develop fundamental theory to improve RF tracking accuracy within highly reverberant urban environments. Dr. Tingley recently completed a study of sensing and tracking systems for the intelligence community in which he managed a team that developed a 5-year technology and product roadmap and is presently executing several development follow-on efforts. Prior to joining Draper Laboratory he was with Bose Corporation, where as lead research engineer he invented, implemented, and patented one of the first (in 1990) software-defined radios developed specifically for commercial application.

Appendix B

Meetings and Participating Organizations

MEETING 1
JUNE 3-4, 2008
WASHINGTON, D.C.

Army Science Board; Defense Advanced Research Projects Agency; Defense Science Board; L-3 Linkabit; U.S. Special Operations Command

MEETING 2
JULY 9-10, 2008
WASHINGTON, D.C.

Harris Corporation; Lockheed Martin Advanced Technology Laboratory; Office of Naval Research; Rockwell Collins; Thales Communications; U.S. Naval EOD Technology Division; U.S. Special Operations Command

MEETING 3
AUGUST 6-7, 2008
WASHINGTON, D.C.

General Dynamics C4 Systems; MITRE Corporation; Ticom Geomatics, Inc.; U.S. Special Operations Command

MEETING 4
OCTOBER 1-2, 2008
WOODS HOLE, MASSACHUSETTS

Writing meeting.

MEETING 5
DECEMBER 2-3, 2008
IRVINE, CALIFORNIA

Writing meeting.

Appendix C
Selected Findings and Recommendations from Previous National Research Council Reports Related to Power and Energy Sources

MEETING THE ENERGY NEEDS OF FUTURE WARRIORS (2004)

Recommendation 5. The Army should refine duty-cycle estimates for the Land Warrior suite of electronics so as to enable the development of high-fidelity models incorporating soldier usage patterns and other details of interactions between power sources and soldier electronics. These estimates are essential for developing smart hybrid systems that can react to the environment for the future LW as well as for developing energy-efficient systems to meet unforeseen Army mission requirements.

Recommendation 10. The Army should make energy efficiency a first-order design parameter whenever specifying system performance parameters in its contracts. It should provide monetary incentives as needed to reduce power demand in all its procurements for soldier electronics, especially for communications.

Recommendation 11. The Army should aim for a future soldier system capable of no more than 2-W average power, 5-W peak power. Achieving this will free the soldier from worries about power shortages on the battlefield and greatly enhance combat effectiveness.

Recommendation 12. The Army should develop a modeling capability for soldier equipment that includes power sources and also enables detailed simulation, verification, and analysis of power requirements for given operational parameters.

ENERGY-EFFICIENT TECHNOLOGIES FOR THE DISMOUNTED SOLDIER (1997)

Recommendation 1a. Army leadership must emphasize the importance of reducing energy demand to achieve energy sufficiency for future dismounted soldiers. Meeting near- and far-term needs will require major changes in Army thinking. Paradigm shifts in energy strategy, system design, and the use of commercial technology are absolutely essential to avert a crisis. The new paradigms must be translated into top-down initiatives. Essential reforms include changes in the following areas:

- Energy Strategy. The Army must focus on energy consuming systems as well as on energy supplying systems.
- System Design for Efficiency. The Army must emphasize system integration at all levels so that the entire system can be optimized for energy efficiency. For example, modular hardware designs with dedicated processors are more energy-efficient than general-purpose computers. And communications architectures must be designed to distribute energy consuming components (sensors and processors) where they can most easily be served by local power sources.
- Use of Commercial Technology. Army systems must be closely coupled to the technologies used in commercial products. The Army must be fully capable of incorporating the most recent data-processing and communications technology into its systems.

Recommendation 1b. The Army should accelerate the development and insertion of enhancements to the Land Warrior system, focusing on improvements to the computer/radio subsystem, because the estimated power requirements for communications and computing functions in Land Warrior are clearly excessive.

Recommendation 1c. The Army Acquisition Executive should make energy efficiency a priority consideration in evaluating contractor performance in future procurements of electronics for the dismounted soldier.

Recommendation 2a. To achieve energy sufficiency, the Army should set research objectives that focus on energy-efficient technologies. Energy efficiency is the key to success for the Army After Next.

Recommendation 2b. The Army should support use of computer-aided design tools for systems and integrated circuits specifically optimized for low power performance. If the necessary design tools are not available commercially, the Army should support its own development programs, perhaps in conjunction with related DARPA efforts. Army contractors for electronic systems should be required to use energy-optimizing design tools.

Recommendation 2c. The Army should support the development of mission-specific software for dismounted soldier systems. General-purpose software is wasteful and not energy-efficient.

Recommendation 2d. The Army should support the development and use of low power software, in which each instruction is written or compiled to minimize power requirements. New tools may be required for specific military applications.

Recommendation 2e. The Army should use dedicated electronic circuits wherever possible to minimize power requirements. Application-specific integrated circuit (ASIC) technology can achieve the efficiencies of custom circuits and hardware and still be cost effective.

Recommendation 2f. The Army should establish and enforce standards of awareness and discipline for energy consumption in dismounted soldier operations.

Recommendation 4a. The Army should refine its requirements for high-resolution images and video communications to the minimum necessary to meet battlefield needs.

Recommendations 4b. The Army should minimize wireless data transmissions by reducing the time required to convey a given amount of information. Relevant technologies include speech and image compression, database caching, and information science technologies that reduce, eliminate, or automate the energy inefficient natural language (read message) transmissions that are currently used.

Recommendation 4c. The Army should adapt the hierarchical network architecture of cellular telephones to create a "virtual peer-to-peer" network, which would improve the distribution of computational resources while taking advantage of commercial cellular technologies.

Recommendation 4d. The Army should modify and synchronize operational doctrine with emerging systems to minimize soldier transmissions. For example, data collection and reduction should be performed as close to the data collector as possible, and computational components should be distributed across the network of soldier communicators. The Army should exploit energy saving communications protocols, such as the protocols used to alert radio receivers to incoming data in pagers and cellular phones. Other commercial techniques should be incorporated doctrinally to reduce or eliminate the operational demands on transmit energy.

Recommendation 4e. The Army should study alternatives for the military network design to optimize power consumption. For example, it should investigate the use of commercial low-orbit satellite systems and unmanned aerial vehicles as relatively energy-efficient alternatives that may also provide high-bandwidth capabilities.

REFERENCES

NRC (National Research Council). 1997. Energy-Efficient Technologies for the Dismounted Soldier. Washington, D.C.: National Academy Press. Available from http://www.nap.edu/catalog.php?record_id=5905.

NRC. 2004. Meeting the Energy Needs of Future Warriors. Washington, D.C.: The National Academies Press. Available from http://www.nap.edu/catalog.php?record_id=11065.